USS Massachusetts On Deck

Written by David Doyle

Squadron/Signal Publications

Art by Don Greer

(Front Cover) While the *Massachusetts* did engage other ships in gunfire duels, most of the battleship's service life was spent conducting shore bombardment and providing antiaircraft support for carriers in the Pacific. Here *Massachusetts* is seen from the USS *Bennington,* while screening for that vessel in April 1945.

(Back Cover) The *Massachusetts* leaves Puget Sound after her final overhaul in January 1946. Much of the antiaircraft armament the ship had carried during WWII has been removed to reduce weight and manpower requirements. The battleship would be decommissioned 14 months later and placed in the reserve fleet.

About the On Deck Series®

The On Deck® series is about the details of specific military ships using color and black-and-white archival photographs and photographs of in-service, preserved, and restored equipment. *On Deck®* titles are picture books devoted to warships.

Hardcover ISBN 978-0-89747-699-7
Softcover ISBN 978-0-89747-700-0

Proudly printed in the U.S.A.
Copyright 2012 Squadron/Signal Publications
1115 Crowley Drive, Carrollton, TX 75006-1312 U.S.A.
www.SquadronSignalPublications.com

All rights reserved. No part of this publication may be reproduced, stored in a retrieval system, or transmitted in any form by means electrical, mechanical, or otherwise, without written permission of the publisher.

Military/Combat Photographs and Snapshots

If you have any photos of aircraft, armor, soldiers, or ships of any nation, particularly wartime snapshots, please share them with us and help make Squadron/Signal's books all the more interesting and complete in the future. Any photograph sent to us will be copied and returned. Electronic images are preferred. The donor will be fully credited for any photos used. Please send them to the address above.

(Title Page) Since 1965, the USS *Massachusetts* (BB-59) has made her home at Battleship Cove in Fall River, Massachusetts. The ship has been designated a National Historic Landmark and hosts large numbers of visitors year around.

Acknowledgments

This book would not have been possible without the generous help of Chris Nardi of the USS *Massachusetts* and my friends Tom Kailbourn, Tracy White, and Scott Taylor. As always, my wonderful wife Denise took notes, scanned photographs, and accompanied me on some very hot days while photographing this storied vessel. Special thanks to the professional editorial team at Squadron Signal Publications, without whose careful attention to detail and countless hours of work this book could never have been created.

Introduction

The Battleship *Massachusetts* was built by the Fore River Shipyard of the Bethlehem Steel Corporation in Quincy, Massachusetts. A *South Dakota* class battleship, she was the fourth US Navy ship to be named for the Commonwealth of Massachusetts.

She was launched on 23 September 1941 and following her fitting out was delivered to the Boston Navy Yard in April 1942 and commissioned the following month. After a brief shakedown period *Massachusetts* sailed from Casco Bay, Maine on 24 October 1942 bound for North Africa, serving as the flagship for Task Group 34.1. While off the city of Casablanca during Operation Torch, *Massachusetts* came under fire from the unfinished French battleship *Jean Bart,* moored at a Casablanca pier, as well as various other French vessels and shore batteries. *Massachusetts*'s 16" main battery answered the French fire, and within sixteen minutes had scored five hits on the French battleship, silencing the *Jean Bart's* guns. Additional *Massachusetts* rounds helped sink two destroyers, two merchant ships, a floating dry-dock, and destroyed an ammunition dump in Casablanca.

Subsequently, *Massachusetts* returned to Boston for refitting before transiting the Panama Canal in February 1943 en route to the Pacific, where she would remain throughout the duration of the war. The battleship saw action in the New Guinea-Solomons area, participating in the invasion of the Gilbert Islands in November 1943, the invasion of the Marshall Islands in January 1944, the powerful carrier strikes against Truk in February 1944, as well as a series of raids against Japanese bases in the Western Pacific and Asia.

After bombarding Ponape Island in May 1944, Battleship *Massachusetts* sailed for Bremerton, Washington for modernization and a well-earned leave for her crew. The break was short-lived, and in September 1944 the ship returned to action in the invasion of the Palau Islands as an escort for the fast carrier task forces.

The invasions of Iwo Jima and Okinawa in 1945 brought the 16" guns of the battleship into action again, as those islands were subjected to thunderous bombardments. In July of 1945 the muzzles of the big guns turned toward the main island of Japan and shelled the Imperial Iron and Steel Works at Kamaishi and a factory at Hamamatsu. Returning to Kamaishi, Battleship *Massachusetts* fired the last American 16" projectile of the war. After V-J day the battleship returned to the United States, soon to enter an extended slumber as part of the reserve fleet, before ultimately being preserved as a memorial.

On her commissioning day, 12 May 1942, USS *Massachusetts* presented this relatively sleek appearance. As the war progressed, however, the ship was continually updated with increasing numbers of antiaircraft guns, radar and radio antennas. (National Archives)

After first seeing combat in the Atlantic, then after spending over a year in combat operations in the Pacific, *Massachusetts* put into Puget Sound Naval Shipyard in June 1944 for overhaul and modernization. She left the yard on 15 July 1944, four days after this photo was taken. (National Archives)

Massachusetts arrived at Puget Sound Naval Shipyard for overhaul in September 1945, after a 5,000-mile, non-stop voyage from Tokyo laden with returning servicemen. After this 22 January 1946 test run following the overhaul, she steamed for the Atlantic and in May 1946 was laid up in the Reserve Fleet. (Naval History and Heritage Command)

The *Massachusetts* slumbered for almost 20 years in the Navy's reserve fleet at Norfolk Navy Yard. As part of the "mothball" fleet, the ship's interior spaces were sealed and dehumidified. Lightweight "cocoons" enclosed the ship's 40mm battery and desiccant materials protected the guns from humidity. (Naval History and Heritage Command)

When the navy was preparing to sell the ship for scrap in 1962, her World War II crew worked with the schoolchildren of Massachusetts to save her as a war memorial. Ultimately the Navy agreed to transfer the battleship to the Commonwealth of Massachusetts. Here she begins the Norfolk-to-Fall River tow. (Naval History and Heritage Command)

With considerable fanfare, the battleship arrived at Fall River, Massachusetts in June of 1965. The ship is preserved today in a configuration similar to that following her 1946 overhaul at Puget Sound Naval Shipyard. Thus the light anti-aircraft weapon battery is greatly reduced from its late-WWII peak. (Naval History and Heritage Command)

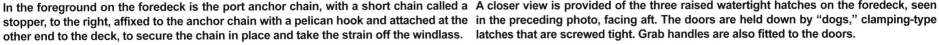

The ship's number on the bow, 59, represents the USS *Massachusetts*' designation, BB-59, signifying it was the 59th contracted battleship of the U.S. Navy. It is white with a black shadow and is a consistent with the post-WWII Measure 13 paint scheme.

In the foreground on the foredeck is the port anchor chain, with a short chain called a stopper, to the right, affixed to the anchor chain with a pelican hook and attached at the other end to the deck, to secure the chain in place and take the strain off the windlass.

At the center of the foredeck is the jackstaff, with two diagonal braces. To each side of the jackstaff are bullnoses: openings in the forecastle through which mooring lines and hawsers were passed. For safety, steel grilles are fitted over the bullnoses.

A closer view is provided of the three raised watertight hatches on the foredeck, seen in the preceding photo, facing aft. The doors are held down by "dogs," clamping-type latches that are screwed tight. Grab handles are also fitted to the doors.

Two single 20mm Oerlikon gun mounts are on the foredeck of USS *Massachusetts*. Each one is partially protected by a curved splinter shield attached to the deck. The gunner received a degree of extra protection from the armored shield fastened to the mount.

The 20mm gun mount on the port side of the foredeck is viewed from a different angle, showing its tripod stand, armored shield, and details of the splinter shield, with its diagonal braces. Aft of the mount is the port hawse pipe; to the left is a watertight hatch.

As viewed from the forward end of the foredeck, the USS *Massachusetts*' two forward 16-inch/45-caliber turrets are dominated by the superstructure and forward fire-control tower rising above them. The forward turret was designated turret one, while the one immediately aft of it was the turret two. During a refit in mid-1944, two quadruple ("quad") 40mm gun mounts were placed in raised tubs on the foredeck, midway between the photographer's vantage point and turret one, but these were removed after the end of the war. Whereas the *Massachusetts* had only six quad 40mm gun mounts in mid-1942, by August 1944 her complement of these mounts had increased to 18.

Atop the 20mm gun's receiver are a drum-shaped 60-round magazine and part of a ring sight. The gun had a theoretical rate of fire of 450 rounds per minute, so the loader kept busy removing empty magazines and installing full ones as fast as possible.

The starboard 20mm gun mount currently on the foredeck lacks a shield and ammunition magazine. Instead of the ring sight, the Mk. 14 computing sight was often mounted on these guns, enabling gunners to better track fast-moving aircraft.

The gunner manipulated the train (or traverse) and elevation of the 20mm gun by strapping himself into shoulder rests on the gun carriage (not present here) and using his body to aim the piece. Grab handles are on the front and rear of the magazine.

The starboard forward 20mm gun and splinter shield are viewed from the center of the foredeck. Numerous 20mm antiaircraft gun mounts were added or deleted throughout World War II, but the two forward mounts on the foredeck survived.

The two hawse pipes on the foredeck (the starboard one is shown), seen in several preceding photos, are where the anchor chains pass down through the bow. When raised, the shanks of the anchors are housed in the hawse pipes.

The port hawse pipe and the protective grilles over it are visible. Light-duty dogs hold down the forward section of grille. Along the deck in the background are several mooring chocks, a mooring bitt, and stanchions for the guard ropes and nets.

The aft portion of the starboard hawse pipe is viewed; the grille over the pipe is intended to prevent falling into the pipe. Two stoppers are fastened with pelican hooks to the anchor chain; the stoppers are attached to the white-painted padeyes on the deck.

The port hawse pipe is seen facing aft. Four metal tabs welded to the rim of the grille serve to engage the dogs, which hold down the grille to the raised coaming of the hawse pipe. A single bitt is incorporated into the rear corner of the hawse pipe coaming.

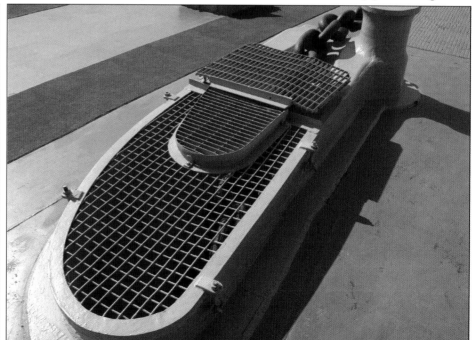

A roller chock is provided on each side of the foredeck. Mooring lines, or hawsers, would pass through this chock. The rollers were intended to reduce chafing on the lines; a severed mooring line could result in dangerous or disastrous circumstances for both ship and crew.

The anchor chains pass through riding chocks, or fairleads, partway between the hawse pipes and the windlasses. Attached to the deck below the anchor chains are chafing plates, to protect the deck from the grinding action of the anchor chains when in use.

The forward part of the foredeck was made from welded steel plates. In the foreground on the deck can be seen the faint contours of where an anchor washboard (a recessed platform for a spare anchor) once sat. This area subsequently was home for a 40mm quad mount, and ultimately was filled in and brought flush with the rest of the deck.

Fastened to the deck between the riding chocks are four padeyes, painted white, two of which are fitted with shackles. These padeyes served to secure stoppers. Aft of the padeyes is the bottom of a ship's anchor, on display here.

The display anchor is viewed facing forward on the foredeck, with the shackle at the top of the stock of the anchor in the foreground. At the other end of the anchor are the flukes. To the left, a good view is provided of a pelican hook on an anchor-chain stopper.

The port anchor chain passes through this riding chock, or fairlead. The view is facing aft. The rear of the visible part of the anchor chain is engaged around the port wildcats. Part of the windlass, the wildcat was a drum that raised and lowered the anchor.

Forward of the port wildcat is this raised watertight hatch. The door hinged to the top of the hatch coaming has a small, circular scuttle with a locking hand wheel. To the front of the hatch is a red fire main. Aft of the hatch is a capstan for operating mooring lines.

To the left of the capstan is the port windlass. The drum of the windlass is called the wildcat and has indentations that engage the anchor chain links, acting similarly to a drive sprocket in raising or lowering the chain. To the far left is the starboard wildcat.

Aft of the wildcats is the breakwater, a low bulwark across the main deck intended to divert waves that crash on the deck from flowing into the area of the forward 16-inch gun turrets. The muzzles of those guns are fitted with tompions to seal off the barrels.

The front facet of the breakwater slants forward at the top. The top of the breakwater is rolled, to afford protection to personnel who might bump into it. Abutting the center rear of the breakwater is a plenum chamber with four circular cover plates.

From each wildcat, the anchor chain is routed down the chain pipe to the chain locker. Aft of each of the wildcats are two hand wheels on pedestals; one controlled the speed and direction of movement of the chain, and the other was for braking the chain.

Stowed to each side of the plenum chamber aft of the breakwater is a paravane, a towed device designed to detonate mines or sever their anchoring cables. The paravane is viewed from the front, showing the wing mounted over the body.

Viewed from the starboard side, the breakwater is stiffened at the rear by triangular braces. On the near side of the paravane is a ventilator, and on the opposite side of the paravane is the plenum chamber. The front of the teak part of the deck is at the right.

To deploy the paravane, a boom would be erected on the side of the hull, and the paravane, tethered to the boom with a cable, would be lowered into the water. Thence, it would move away from the ship, gliding in water.

The foredeck is viewed from behind the starboard side of the breakwater, showing the contrast between the teak deck in the foreground and the steel-plate deck forward of the breakwater. A small ladder with two treads is mounted on the plenum chamber.

When deployed, a fin-type stabilizer was attached to the rear of the body of the paravane. Paravanes were effective in defeating moored mines and in detecting mined seaways, but they also decreased the speed, maneuverability, and endurance of the ship.

Looking across the front of the breakwater from the port side, the top of the port paravane is visible. Beyond the paravane is the top of the plenum chamber. To the far right is the front of turret one, while to the left are the wildcats and windlass controls.

The front of turret one is viewed over the top of the port wildcat (bottom) and the plenum chamber. Black blast bags, or bloomers, are fitted over the barrels of the 16-inch/45-caliber guns to seal the gaps between the gun barrels and the fronts of the turrets.

The blast bags on the front of turret one are seen from the starboard side. Secured to the barrels with clamping bands, the blast bags were fabricated from rubberized fabric and were intentionally loose fitting, to accommodate the elevation and recoil of the guns.

The barrels of the three 16-inch/45-caliber Mk. 6 guns of turret one have three steps, or hoops. The "45-caliber" part of their designation indicates that the length of the barrel was 45 times that of the bore of the gun; hence, the bore was 720 inches long.

The armored housing of the 16-inch turret was called the gun house. This is the starboard side of the gun house of turret one, with turret two looming in the background. The armored face plates are 18 inches thick, while the side plates are 9.5 inches thick.

Projecting from the side of the gun house are hoods for the right trainer's (upper) and right pointer's (lower) telescopes. To the rear of the side of the gun house is the hood for the right side of the optical rangefinder; these hoods were nicknamed "ears."

The right rear side of turret one is viewed facing forward. The rear of the right rangefinder hood is to the right. Below the hood is a ventilator plenum on the side of the gun house. A larger ventilator plenum is prominent on the rear plate of the gun house.

The rear of the gun house of turret one is viewed from a different angle, showing the two ventilator plenums, fabricated from welded steel. The rear plate of the gun house is 12 inches thick. Faintly visible below the turret is a raised watertight hatch.

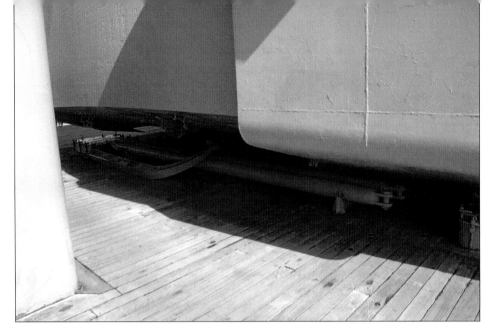

Stowed below the rear overhang of turret one are several tubular struts and a small davit, probably for lowering ammunition or cargo below decks. The hatch providing the crew with access into the gun house is located toward the center of the bottom of the overhang.

Visible on the port side of turret one are the left trainer's and pointer's telescope hoods and the left rangefinder hood. The ring-shaped base the gun house rests on is the top of the barbette, a heavily armored structure extending deep into the hull.

The front of the left hood for turret one's rangefinder is viewed. The rangefinder's objective was protected by a sliding armored shutter, operated from within the gun house. The rangefinder operator's station was near the center rear of the of the gun house.

As observed from the navigating bridge of USS *Massachusetts*, the 16-inch/45-caliber guns of turrets one and two mutely stand watch over the Taunton River. The roof plates of the 16-inch gun houses were fabricated from 7.25-inch armor plate. Multiple lateral tiers of oval-head slotted screws are visible; these secured the plates to the frame of the gun house. Toe rails are fitted around the forward and side edges of the roof. Two periscope heads are present on the roof of turret one, visible on either side of the center gun of turret two.

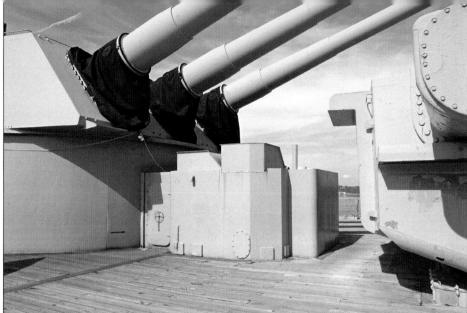

Jutting from the front of the barbette of turret two on the main deck level is a structure designated the deck office, and further designated as compartment A-101L. On the side of that structure adjacent to the barbette is a door with a hand-wheel lock.

Mooring bitts like this one on the starboard side of the main deck serve as anchoring points for heavy lines called hawsers, the opposite ends of which were secured to a wharf or quay. In this case, heavy chains are secured to the bitt and the quay alongside the ship.

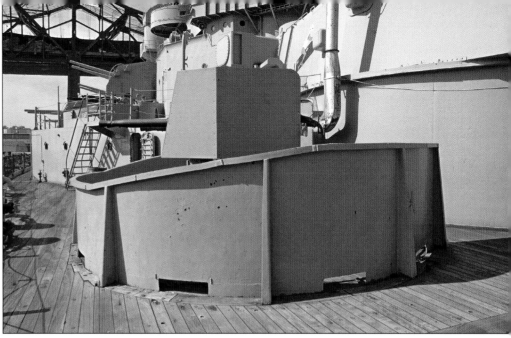

Two quad 40mm antiaircraft guns and gun tubs were installed on the main deck to each side of turret two after the USS *Massachusetts* was commissioned. Shown here is the starboard one. Rising above the gun tub is the left side of the gun shield.

The quad 40mm gun mount shown at left is viewed more closely. The cutouts at the bottom of the gun tub let crewmen kick spent shell casings out of the tub, preventing fouling the mount. The cutouts also allowed water that washed into the tub to escape.

The starboard forward 40mm gun mount is viewed from aft. The 40mm mounts in this part of the main deck were not installed until 1944; prior to that, there was a gallery of 20mm guns behind splinter shields in this area on each side of the main deck.

The same 40mm gun mount shown in the preceding photos is seen facing forward. In addition to the cutouts in the front plate of the shield for the gun barrels, there are cutouts toward the sides for the pointer's and trainer's sights.

The starboard forward quad 40mm gun mount is viewed from the rear. Seats are provided for the pointer, left, who manually elevated the guns, and the trainer, right, who traversed the mount. The platforms were for the loaders and the gun captain.

In a view of the lower front of the quad 40mm gun mount, to the left are the seat and foot rests for the trainer. In the foreground is the train power drive. The curved object at the center is a case discharge chute; three more of these chutes originally were present.

Aft of the forward starboard 40mm gun mount is this angled facet of the first level of the superstructure. The door provides access to a corridor leading to several wardroom staterooms. Aft of the bottom of the ladder is a red-colored fire main.

This photo taken on the starboard side of the main deck around frame 80 shows how, aft of the angled facet of the superstructure, the first level of the superstructure runs parallel to the edge of the main deck. The door in the foreground leads to the flag office.

The section of main deck shown in the preceding photo appears at the bottom of this view of the starboard side of the superstructure. The *South Dakota*-class battleships were designed and constructed under tight size and weight constraints, which included a fairly compact arrangement of a single smokestack (left) abutting the rear of the forward fire-control tower (upper center), as seen here. Also, the secondary battery of twin 5-inch/38-caliber gun mounts was compactly arranged, with five mounts on each side of the superstructure. Four of the starboard 5-inch mounts, omitting the aft mount, are in this view. To the far right is the navigating bridge, above which is the top of the conning tower.

The starboard 5-inch/38-caliber gun mounts are visible here. The 5-inch mounts were assigned numbers, one through ten, forward to aft, with the odd numbers being on the starboard side and even numbers on the port side. Hence, from right to left, mounts one, three, five, seven and nine are present in this photo. There are four Mk. 37 directors on the ship. These controlled the fire of the secondary battery of 5-inch/38-caliber guns. Two of the Mk. 37 directors are visible here, to the upper right and at the upper center. Each of these directors is topped with a Mk. 12 radar antenna with a Mk. 22 "orange peel" altitude-finding radar antenna next to it on the right side. To the far left is part of the boat crane.

This feature, also visible at the lower left in the preceding photo, is a Mk. 51 director (or Mk. 51 fire-control system, abbreviated FCS) protected by a splinter shield. Each Mark 51 director controlled the operation of a quad 40mm antiaircraft mount; in this case, the associated 40mm mount is to the far left. This director is on the main deck at the same level as its associated 40mm gun mount. More often, Mk. 51 directors were situated above and slightly away from their associated gun mounts, in order to give the director operator a better vantage point. Above this director is a stair landing, and above the landing, mounted on the superstructure, a ship's bell is partially visible.

The same Mk. 51 director seen in the preceding view is observed from aft. The part of the director visible above the top of the splinter shield is the front of a Mk. 15 lead-computing gun sight. The operator of the director manipulated the gun sight to track attacking aircraft, and the director computed the amount of lead necessary and electronically transmitted the firing solution to the associated quad 40mm gun mount. If necessary, however, the gun crew could aim and fire the guns. The splinter shield has rolled top and side edges. Above the director is the right side of the shield (as the enclosure of the gun mount is called) of a twin 5-inch/38-caliber gun mount

The Mk. 51 director, viewed from the rear, comprises a Mk. 15 gun sight mounted in a yoke on top of a pedestal attached to the deck. The unit included a gyro which, when caged, was aligned with the gun sight's line of sight. When the operator acquired a target, began tracking it, and uncaged the gyro, the sight transmitted information to a computer, which calculated the necessary lead for the 40mm guns. To track the target, the operator manipulated the bicycle-type handlebars. The two objects protruding to each side of the gun sight are counterweights. Before the Mk. 15 gun sight was introduced, the Mk. 14 gun sight was used in the Mk. 51 director. An auxiliary telescope also could be mounted on the sight.

The right side of the same Mk. 51 director is in view. The right handlebar control curves down from the sight bracket. The mechanism with the round, spoked cover to the front of the director is an air pump, to supply compressed air for driving the gyro.

Visible on the left side of a Mk. 51 director are the left handlebar control and the sides of the Mk. 15 gun sight and the air pump. Extending to the rear from the upper part of the director is the left counterweight. To the far left is the shield of a 40mm gun mount.

The upper part of the Mk. 51 director, including the Mk. 15 computing gun sight, is viewed from the front. This sight was a complex instrument, containing sophisticated optical and gyroscopic mechanisms. On the top half of the sight is the sight window.

The Mk. 15 director is observed from above. The operator stood between the two counterweights. The eyepiece is the dark fixture on the right rear of the sight. On top of the sight are two ring clamps for attaching an auxiliary telescope.

From the landing on the first superstructure deck above the Mk. 15 director portrayed in the preceding series of photos, the starboard side of the main deck aft of the superstructure is in view. In the foreground is a quad 40mm gun mount and tub.

The rear of the 40mm gun mount and its gun tub seen in the preceding photo are viewed from the main deck, facing aft. The splinter shield was manufactured from fairly thin steel. The four curved, vertical objects on the rear of the mount are spent-casing chutes.

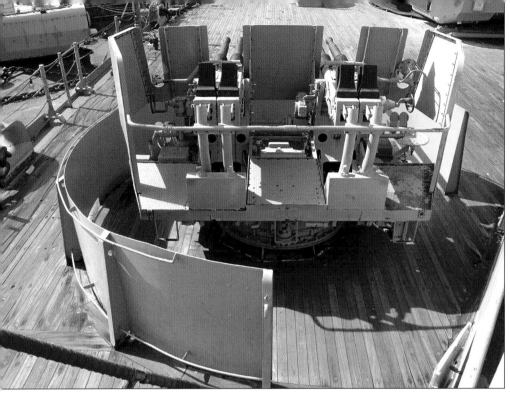 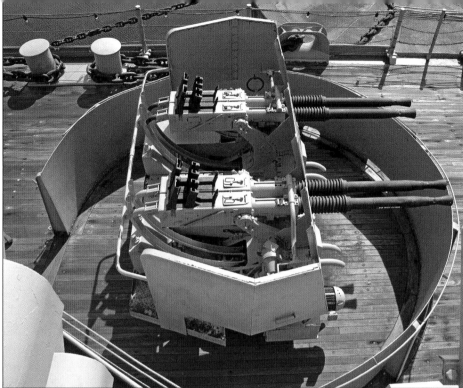

The 40mm gun mount and tub adjacent to the starboard aft corner of the superstructure are observed from above. Steps are present on the rear of the loaders' platform, and the guard rail at the rear of the mount was for the safety of the loaders.

The inner sides of the tubs for the quad 40mm guns on the USS *Massachusetts,* as well as other U.S. navy ships in World War II, were covered with racks for stowing ready clips of 40mm ammunition. These racks were removed from the gun tubs after the war.

The *Massachusetts* wore the Measure 22 camouflage scheme shown here from October 1942 until January 1947. This scheme featured Navy Blue on the lower hull, up to the lowest point of sheer. Above that, the vertical surfaces were Haze Gray, and the decks were Deck Blue. In January 1947 she was repainted into the overall Haze Gray scheme known as Measure 13 – the scheme she wears to this day. From her May 1942 commissioning until October 1942 she wore camouflage Measure 12, a mottled Navy Blue and Ocean Gray hull, as seen at lower left of page 3 of this book.

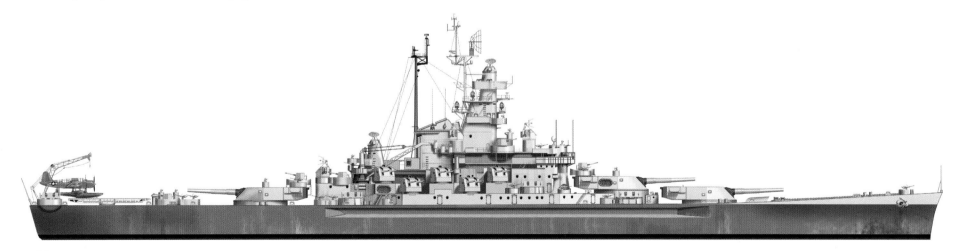

The quad 40mm gun mount viewed in the preceding series of photos is observed from its aft quarter. Above the guns are a ship's bell and five-inch gun mounts numbers 7 and 9. To the left is the aft section of the starboard side of the superstructure. Within this part of the superstructure on the main-deck level was crew's berthing. During the course of World War II, in the left foreground there were several different configurations of 20mm antiaircraft cannons behind a splinter shield, but these were subsequently removed. To the upper left is the left rangefinder housing on turret three. At the top, both the forward (right) and aft (left) Mk. 38 primary-battery directors are visible. These directors, nicknamed, respectively, Spot 1 and Spot 2, controlled the 16-inch/45-caliber guns.

Turret three is observed from the starboard side of the main deck, with the aft Mk. 37 and Mk. 38 directors looming to the right. On display on the side of the turret are a 16-inch projectile and powder charge. Atop the gun house is a quad 40mm gun mount and a Mk. 51 director.

Poised on a cylindrical base atop the rear point of the superstructure of USS *Massachusetts* is the aft Mk. 37 director. Inside the four Mk. 37 directors on the ship, operators, visually and with radar, spotted and tracked aerial or surface targets and communicated that information to a plotting room below decks, where that information was processed into firing solutions for the 5-inch guns. These directors could also control the 40mm antiaircraft guns and, in emergencies, the 16-inch main battery. The director's armored housing enclosed an optical rangefinder and spotting telescopes, and mounted on top were radar antennas: in this case, a Mk. 12 antenna with a Mk. 22 "orange peel" antenna to the side.

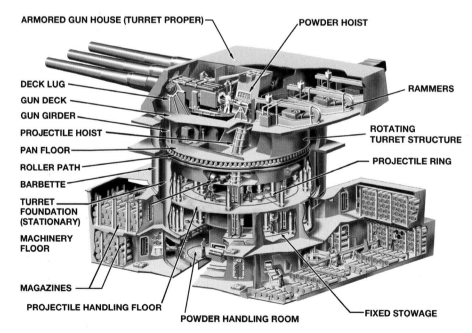

This view of the aft part of turret three from the starboard side provides a sense of the clearance between the rear of the gun house and the superstructure. Originally, a gallery of three 20mm antiaircraft guns was positioned within a splinter shield on the roof of turret three, but during a refitting during World War II, the quad 40mm mount and gun tub were installed atop the turret, and a 20mm gun was mounted on each side of the 40mm mount within a lower, semicircular splinter shield, seen here next to the 40mm gun tub. The 20mm guns are no longer present within those splinter shields. Below the splinter shield for the 20mm gun is the left rangefinder hood.

A closer view is provided of the quad 40mm gun mount and gun tub atop turret three, showing the ladder rungs on the front of the tub and the handrail around the top of it. Details of the Mk. 37 director and its radar antenna array are also visible.

The right 20mm gun station atop turret three is shown. In the foreground is an ammunition locker; on the forward side of the splinter shield is another ammunition locker. Extending below the gun station is the right rangefinder hood.

The left 20mm gun station and splinter shield on top of turret three are viewed facing the stern of the *Massachusetts*. The round mounting plate for the 20mm gun is at the center of the position. To the right is the gun tub of the quad 40mm gun mount.

The 40mm guns in the mount on turret three are visible above the splinter shield. Many of their parts and components have been stripped, including the spent-casing deflectors at the rears of the guns, but the automatic loader chutes atop the guns are still present.

The curved parts below each pair of 40mm guns are elevating arcs. The vertical pole in the right foreground is a davit for hauling ammunition and other materials up to the gun tub. Rising in the background is the aircraft-handling crane on the fantail.

The rear of the 40mm gun tub on top of turret three is viewed close-up. To the upper left are the brackets for the davit. Parts of the foot rail running along the lower part of the tub are missing. To the right, an ammunition locker rests atop a ventilator.

With the left front corner of turret three in the foreground, the 40mm gun tub atop the turret appears to the far right. At the center is the tub for the Mk. 51 director associated with the 40mm gun mount directly behind it. The sight is missing from the director.

The blast bags of the center and left 16-inch/45-caliber guns of turret three are viewed. As built, there was a ladder on each side of the 16-inch guns on the front plate of each turret, but these ladders were removed at some point after January 1946.

On each side of the main deck abeam turret three is a warping winch, used along with hawsers in docking the ship. Each winch has two drums, one on each side of the gearbox, for hauling on the hawsers. The winch motor is forward of the gearbox.

The bulwark seen at a distance in the preceding view is shown close-up. Instead of the original 20mm antiaircraft guns, two other devices, including what appears to be a Mk. 15 director to the left, are on display in this area now. To the left is the aircraft crane.

The starboard warping winch is viewed facing toward the stern of the ship. Directly aft of the winch are a splinter shield that originally enclosed several 20mm antiaircraft guns. Just aft of that shield is a quad 40mm mount that was a retrofit during World War II.

On the starboard side of the main deck just aft of the splinter shield shown in the preceding photo is a quad 40mm gun mount in a gun tub. The sloping top of the gun tub allowed the guns to depress to fire on enemy planes approaching low off the beam.

Turret three, with its 16-inch/45-caliber guns elevated to different angles, is viewed from the after part of the main deck. Like the gun houses of the other 16-inch main-battery turrets, the armor on turret three's gun house is 18 inches thick on the front, 9.5 inches thick on the sides, 12 inches thick on the rear plate, and 7.25 inches thick on the roof. Each of these guns could sustain a rate of fire of about two rounds per minute, with a maximum range of approximately 37,000 yards, or 21 miles. During one of the rare instances where a U.S. battleship exchanged fire with an enemy battleship during World War II, the *Massachusetts* registered five 16-inch hits on the Vichy French battleship *Jean Bart* at Casablanca in November 1942, causing significant damage.

The aft port section of the superstructure of USS *Massachusetts* is viewed from the main deck to the side of turret three. To the left are twin 5-inch/38-caliber gun mounts numbers 8 and 10. Forward of the Mk. 38 director is the mainmast, which replaced the smaller original mainmast during a 1945–46 refit. An SR-A secondary air-search radar antenna is installed near the top of the mainmast. Farther forward is the foremast, which carries an SK-3 air-search radar antenna. To the front of the mainmast is the forward fire-control tower, atop which is the forward Mk. 38 primary-battery director. To the left of that tower is the center port Mk. 37 secondary-battery director, with its suite of Mk. 12 and Mk. 22 radar antennas visible on top.

In a view from the centerline of the main deck facing forward, turret three and the superstructure loom in the background. Several hatchways leading belowdecks are in view. The low structure with the ladder up the side is a large ventilator plenum.

The forward side of the ventilator plenum is viewed, with the aircraft-handling crane on the fantail directly aft of it. To the far left is a quad 40mm gun mount; the Mk. 51 director associated with it is in the small tower abutting the side of the ventilator plenum.

As viewed facing aft on the port side of turret three, features from left to right are: Mk. 51 director; ventilator plenum (behind screen); Mk. 51 director; quad 40mm gun mount and tub; 20mm antiaircraft guns behind splinter shield; and port warping winch.

Lying along the port edge of the main deck abeam turret three are two kingposts for boat booms. When the kingposts were erected on the hull, cables running from their tops supported the boat booms, to which boats were moored when the ship was at anchor.

The kingposts for the boat booms are viewed from a different angle. Standing next to the kingpost on the right is a davit. On the edge of the deck next to the kingpost to the left is a chock, through which mooring hawsers would pass. In the foreground is a ventilator.

The port warping winch on the main deck abaft turret three is viewed facing forward. Along the edge of the deck in the background are the davit and the two kingposts for boat booms seen in the two preceding photos. To the right is the side of turret three.

The port warping winch is viewed from the side, showing the spoked design of the drum. Two small lifting eyes are mounted on top of the gearbox, and one lifting eye is visible atop the motor.

The port side of the main deck aft of the superstructure is viewed from above, showing a hatchway and a ventilator in the foreground, the port warping winch in the middle distance, and the 20mm gun gallery and the 40mm gun tub in the background.

This 20mm antiaircraft gun is on a pedestal mount behind a splinter shield on the port side of the main deck, aft of turret three. The shield comprising two armor plates that originally was fitted on the gun mount is no longer present. Shoulder rests are present, however.

The hand wheels on the pedestals were for raising or lowering the gun cradles, thus placing the guns at an optimal height for the gunner as he maneuvered himself to elevate or depress the piece. Above the guns are the brackets and eyepieces of the ring sights.

Aft of the 20mm guns on the port side of the main deck is this quad 40mm gun mount and tub, with an interpretative placard on the tub. To the right is the Mk. 51 director associated with this 40mm gun mount, situated in a tub atop a pedestal.

From the left side of the quad 40mm gun mount, the elevated barrels virtually disappear against the bridge in the background. The pointer's footrests are visible below the side of the shield; the right footrest doubled as a firing pedal.

The Mk. 51 director associated with the quad 40mm gun mount in the preceding photo is viewed facing aft. The director lacks a sight. The white object on the side of the tub is a Japanese aircraft recognition chart. Fire hoses are stowed on the pedestal of the director.

The same Mk. 51 director seen in the preceding photo is viewed in relationship with the 40mm gun mount it controlled. The pedestal the director and tub are mounted on gave the operator of the director a better field of view over the adjacent quad 40mm gun mount.

As viewed toward the port side of the *Massachusetts*, the tub of the Mk. 51 director displays another Japanese aircraft identification chart on its inboard side. Above the tub, the top of the director mount, including the counterbalances, is visible.

In a view from the rear of the superstructure, the port Mk. 51 director (left) and quad 40mm gun mount (right) featured in the preceding series of photos are observed from above. In the background is the port fantail quad 40mm gun mount and tub.

Mooring bitts, two solid steel posts mounted on a base plate, are found at intervals along the edge of the main deck. Mooring lines were made fast to the bitts to secure the ship to a dock or quay. This bitt is abeam turret three on the port side of the deck.

The fantail, the aft part of the main deck, is clad with steel plate. Flanking the aircraft-handling crane at the center are two quad 40mm gun mounts placed in gun tubs, and two Mk. 51 director tubs positioned on pedestals. Two mooring bitts are to the right.

The mooring bitts were secured securely to the structure of the ship, to withstand considerable stresses. Where a fixture is secured to the deck, such as these bitts, there is a teak frame around it, and the fore-and-aft decking is cut to fit around the frame.

The fantail is observed from the port side. During World War II, the port aircraft catapult would have occupied much of the space in the right side of the photo. The two catapults and their foundations were removed while the ship was in mothballs after the war.

A temporary screen hides most of turret three in this view of the superstructure from a position on the fantail between the two aft Mk. 51 directors. To each side of the screen are other Mk. 51 directors and 40mm gun tubs.

Although the two aircraft catapults are missing from the USS *Massachusetts*, the aircraft-handling crane is still in place. It is observed from the starboard side of the stern. Flanking the crane are the two aft 40mm gun mounts and their gun tubs.

The lower part of the aircraft-handling crane is observed facing aft. The heel, or bottom of the crane boom, is mounted with pins to a base, which in turn rotates with the boom on the stand attached to the deck. The placard is not original equipment.

To the aft starboard side of the aircraft crane are controls for the crane. From left to right, they are: the hoisting-gear operating hand wheel; emergency brake hand lever; speed indicator and electrical control; and turning-gear operating hand wheel.

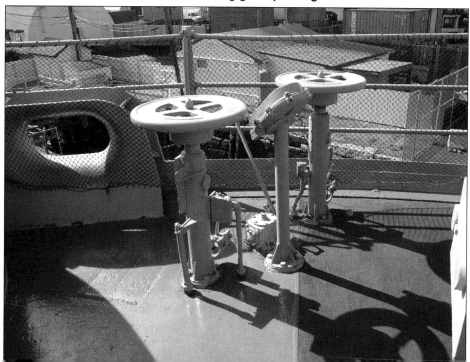

The base of the aircraft-handling crane is observed facing aft, showing the pockets in the top of the base where the heel of the crane is secured. Below the crane on the second deck is the crane's turning gear, while on the third deck is the hoisting gear.

The lower part of the aircraft-handling crane is viewed from its right side. Straddling the sheave at the center is the A-frame. Attached to the top of the A-frame is a tie rod, the bottom of which is pinned to the arm extending to the rear of the base of the crane.

The boom of the crane is seen from its right side. Several spoked sheaves are visible on the crane, as well as two spotlights. When conditions necessitated it, such as in a severe storm, it was possible to lower the crane boom so that it lay flat on the deck.

The base of the crane and the stand below it are observed from the left side, showing the points where the boom (left) and A-frame (right) are pinned to the base. To the upper right is the lower part of the tie rod that connects to the top of the A-frame.

The aircraft-handling crane is viewed from the front, with a Mk. 51 director tub located to each side and a mooring chain running across the deck at the bottom of the photograph. The crane was designed to rotate at a rate of one-half revolution per minute when bearing a full working load. Faintly visible running up the rear of the crane is a ladder, to give riggers access to the upper part of the boom. A removable brace attached to the front of the crane helps stabilize the crane. At least one photo of the USS *Massachusetts* during her time in commission shows such a brace installed.

The two tubs for the Mk. 51 directors associated with the two quad 40mm gun mounts on the fantail are viewed facing forward. Oddly, the starboard tub has its rounded side and access ladder facing aft, while the port tub has its rounded side facing forward.

The close proximity of the port fantail Mk. 51 director tub to the gun tub of the port quad 40mm gun mount is illustrated. Both fantail director tubs currently have had the actual directors replaced by electrical equipment lockers, visible above the tops of the tubs.

The quad 40mm gun mount and tub on the starboard side of the fantail is viewed facing aft. An informational placard is attached to the gun tub. To the upper right is the top of the A-frame of the aircraft-handling crane.

Spent casings were ejected through the ports at the rears of the guns. Deflectors at the rears of the guns guided the casings onto the curved spent casing chutes. Then, the spent casings slid down the chutes and emptied out to the front of the guns.

A view through the guard rails at the left rear of the mount provides a glimpse of the quad 40mm antiaircraft guns. The dark-colored devices on top of the receivers of the guns are the automatic loaders, into which four-round clips of ammunition were inserted.

The tops of all four spent-casing chutes are visible to the rear of the quadruple 40mm guns. The steel used in the splinter shields and gun tubs were relatively thin gauge, intended mainly to protect the gun crews from splinters: flying bits of metal.

The 40mm gun mount and tub on the port side of the fantail are similar in layout to those on the starboard side of the fantail. The "do not climb" stenciling on the gun barrel jackets are non-original markings.

Features at the lower front of the quad 40mm gun mount are displayed. Hinged covers, secured by dogs, provided a degree of protection to the drive mechanisms at the front of the mount. Below those covers are the outlets of the four spent-casing chutes.

The scene is along the starboard side of this main deck, where an assortment of davits, guard-rope stanchions, and posts present a busy appearance. In the background are several 5-inch/38-caliber gun mounts as well as a quad 40mm gun mount.

At the stern, a mooring chain passes through a chock equipped with a hinged retainer plate on top. Adjacent to the chock is the gun tub of the port fantail quad 40mm gun mount. The opening at the bottom of the gun tub allows water to flow out.

When the ship was completed, and throughout WWII, she carried an aircraft catapult on either side of the fantail. These were removed during the 1950s, as efforts were made to keep *Massachusetts* and other battleships up to date even though they were laid up in the reserve fleet. This photograph was taken 26 January 1948 to illustrate damage to the boat boom caused by another vessel colliding with the battleship (Norfolk Navy Yard).

The *Massachusetts* carried a number of different OS2U-1 and OS2U-3 scouting planes during her career. The initial complement was three aircraft, with one stowed on each of the two catapults and the third stowed on the deck between them. However, the number of aircraft assigned were sometimes less. They were launched via the catapults and recovered by landing in the sea, being hoisted aboard by the stern crane. This particular aircraft was carried during Operation Torch.

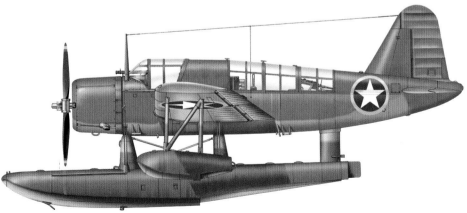

At the stern of the USS *Massachusetts* is the flag staff, from which the national ensign was flown when the ship was at anchor or docked between the hours of 0800 and sunset. The national colors were to be transferred from the gaff to the flag staff at the moment the anchor was dropped. The flag staff is set at a rearward angle so that it hangs over the stern. The disc-shaped object at the top of the flag staff is the truck.

The USS *Massachusetts* is viewed from a distance at her berth in Battleship Cove, Fall River, Massachusetts. Visible here is a very pronounced feature of *South Dakota*-class battleships: a recess in the side of the hull just below the main deck, extending from forward of turret two to the rear of the superstructure. The purpose of the recess was to form a sort of shelf along the top of the ship's side tankage (i.e., at the bottom of the recess), to provide locations for the fuel-oil filler plates at the tops of the tanks. The tanks built into the side of the hull included air voids and fuel tanks. In addition to providing storage space for fuel, this system of voids and tanks provided anti-torpedo protection to the most vital internal spaces of the hull.

The method of attaching the blast bags, or bloomers, to the 16-inch guns is illustrated in this view of the front left corner of turret one. A band-type clamp secures the bag to the gun, and the rear of the bag is attached to an angle-iron bracket surrounding the gun port. The rubber-impregnated canvas bloomers do not have an indefinite life span, and are subject to deterioration from sunlight and the elements. Fortunately for the *Massachusetts* and other preserved battleships, the Navy's reactivation of the *Iowa*-class battleships during the Reagan administration and their subsequent deactivation in 1990-1991 created fresh surplus supplies of these components.

This is the quad 40mm gun mount and gun tub on the main deck adjacent to the forward port corner of the superstructure. The guns are trained outboard, and the right side panel of the splinter shield is in view. Conical flash supressors are on the muzzle of each gun.

The quad 40mm gun mount on the main deck adjacent to the forward port corner of the superstructure is observed facing turret two and the superstructure. The panels of the gun's shield are attached to the inner frame of the shield with slotted screws.

Number of Light Anti-Aircraft Armament Mounts Aboard *Massachusetts*

Date	40mm Quad	20mm Single	20mm Twin
May 1942	6	12	N/A
November 1942	6	35	N/A
January 1943	10	50	N/A
February 1943	12	61	N/A
June 1944	16	52	1
August 1944	18	38	1
August 1945	15	22	8

An experimental 20mm quad mount was aboard from June 1944 through October 1945.

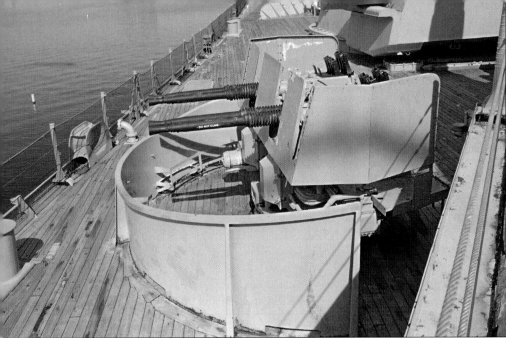

The same quad 40mm gun mount seen in the preceding photo is viewed from the superstructure deck, giving a good idea of the amount of working space the gun crew had available. Recoil springs at the rear of the barrels are visible.

The quad 40mm gun mount on the main deck adjacent to the forward port corner of the superstructure is seen from aft. The curved sections of teak planking on the deck in the foreground indicate the location of a splinter shield for a 20mm gun that was removed.

From about frames 69 to 99, the first level of the superstructure runs parallel to the edge of the main deck, leaving a narrow walkway along that deck. This view illustrates a portion of the port side of that part of the superstructure, showing several of the doors and portholes. A semicircular rain deflector is mounted above the porthole in the foreground. The doors open outward, making for more constriction on the walkway when they were opened. Within view on the superstructure deck above are several of the 5-inch/38-caliber gun mounts.

Alongside the port aft end of the superstructure on the main deck is a quad 40mm gun positioned inside a tub, corresponding with a similar gun mount on the opposite side of the main deck. This gun mount lacks the guard rails, the left gun's spent-casing deflector is missing, and the spent-casing chutes are no longer present. However, other key components are intact, including the ring sights and their brackets as well as the automatic loader assemblies on top of the receivers of the guns. In the left background is turret three.

The same 40mm gun mount seen in the preceding photograph is viewed from above. The ready racks for 40mm ammunition that covered the inside of the gun tub during World War II were long since removed.

To the right is the 40mm gun mount adjacent to the aft port side of the superstructure. Farther aft along the edge of the main deck is a davit, two kingposts for boat booms, and a mooring bitt. To the far left, partially hidden by a safety rail, is the port warping winch

The starboard side of USS *Massachusetts* is observed. To the right is the ship's mooring quay. The features of the ship aft of the 40mm gun tub atop turret three are hidden by several other ships and boats on display at Battleship Cove, including the fleet submarine USS *Lionfish* (SS-298). The recess along the top of the hull is quite noticeable. Atop the foremast, located aft of the forward fire-control tower atop the superstructure, is a dish-shaped SK-3 air-search radar antenna. Aft of the foremast is the ship's sole smokestack. The single smokestack was a design consideration that grew out of the necessity to make the ship's above-decks structures as compact as possible, due to weight and size concerns.

The 16-inch/45-caliber guns of turret two are viewed from the starboard side of the main deck. Details of the right rangefinder hood of turret one are visible to the right. On the outboard side of the hood is an access plate, attached with hex screws to the hood.

The blast bags and front plate of turret two are observed up-close. The armor of the frontal plates of the turrets was the thickest on the ship, at 18 inches. Not far behind was the armor of the conning tower, which was 16 inches thick on the sides.

On the right side of turret two is a recreation of a cartoon painted on the turret during World War II, showing George the Gremlin riding a 16-inch projectile with Japanese and German flags painted on it. At the top left are the bridge and top of the conning tower.

More of the right side of turret two is observed from the superstructure deck. Below the rangefinder hood is a ventilator plenum. Another ventilator plenum is on the rear of the turret; the top of the ventilator projects out from the trunk of the ventilator.

A view of the side of turret two taken from alongside the rear overhang shows toward the front the hoods for the right pointer's and trainer's telescopes. The pipe attached by brackets to the bottom of the gun house is a foot rail that was retrofitted at some point.

The hatch to the gun house of turret two is in the bottom of the overhang. The platform under the hatch as well as the foot rails enabled personnel to safely exit the turret when it was trained to the side; otherwise, it was a 10-foot fall to the main deck.

Facing forward from the front starboard side of the superstructure deck, turret two is to the left, a 40mm gun mount is at the center, turret one is in the middle distance, and the foredeck is in the far distance. The stack next to turret two is not original equipment.

In a view of the starboard side of the superstructure deck facing aft, the right rangefinder hood of turret two is to the right, above which part of the bridge is visible. Between the quad 40mm gun mount and the 5-inch/38-caliber gun mount is a Mk. 51 director.

The forward part of the superstructure of the USS *Massachusetts* is observed from the starboard side of the main deck, with part of turret two to the right and the bridge immediately above and aft of the turret. Rising above the rear of the bridge is the top of the conning tower, the heavily armored nerve center of the ship during battle. Just aft of the conning tower is the forward fire-control tower, complete with a yardarm and topped with the forward Mk. 37 secondary-battery director, while above and aft of that director is the forward Mk. 38 primary-battery director, "Spot One."

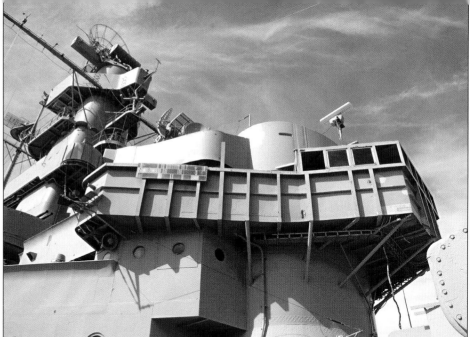

The board on the rear part of the bridge is painted with campaign ribbons representing the campaigns in which the USS *Massachusetts* saw action.

At the superstructure-deck level, the angled facet on the forward starboard side of the superstructure aft of the conning tower (visible to the right) has two doors on the first level, along with exposed electrical cables. The upper level has two portholes.

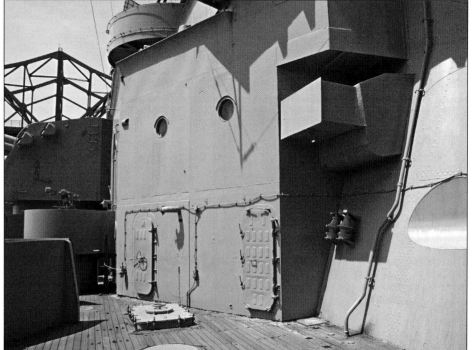

As viewed from the starboard side, the conning tower has a pronounced taper from the level of the superstructure deck up to the level below the navigating bridge deck. Viewed from above, the upper part of the conning tower has an oblong shape, wider from side to side than from front to back. The navigating bridge was remodeled several times during and once after World War II. A 1946 refitting included the enlarging of the bridge, installation of support struts underneath the bridge, and the addition of the enclosure at the center part of the bridge.

Adjacent to the angled facet of the port side of the superstructure on the superstructure deck is a quad 40mm gun and tub. The Mk. 51 director that controlled this gun mount is partly visible between the 40mm gun tub and the nearest 5-inch gun mount.

The Mk. 51 director on the superstructure deck adjacent to the forward port facet of the superstructure is viewed. The gun sight is not installed on the director, but the cradle and the two counterweights are visible. A spoked cover is present on one of the portholes.

The Mk. 51 director and its tub are viewed from the rear, showing its slightly higher position than that of the 40mm gun mount it controlled. Present on the tub are a cleat, a tie-down eye, and slot-shaped scuttles to allow water to escape from the tub.

Five-inch/38-caliber mount number one, on the starboard side of the USS *Massachusetts*, is observed from below. The shield of the mount was formed of two-inch-thick steel armor. The shields are about 15 feet wide, 10 feet high, and 16 feet from front to rear.

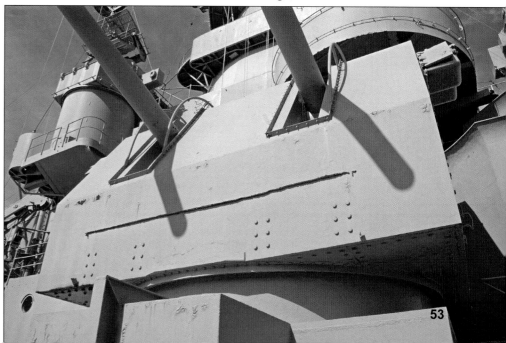

As viewed from the right side of 5-inch/38-caliber gun mount one, the protective hood for the trainer's sight protrudes above the access door. On the rear of the shield is an oblong crew access door with three rungs mounted on the armor below the door.

On the left side of the 5-inch/38-caliber shield are two armored hoods for sights. The forward hood is for the trainer's sight, while the hood behind it is for the sight checker, who ensured that the sights of the gun mount were laid on the designated target.

This space is on the superstructure deck between the superstructure (right) and 5-inch gun mount number three. The assembly on the side of the superstructure that is shaped like a quarter of a cylinder is a ventilator hood.

Five-inch gun mount five rests on top of a raised structure that serves as the gun mount's upper ammunition-handling room. The view is facing aft. To the top left is the roof-hatch operating mechanism on the rear of the shield of 5-inch gun mount number three.

The number-five gun mount is viewed from farther back, showing the typical sight hoods for the trainer and the sight checker. An access door is on the side of the shield below the trainer's sight hood. Below the shield of the turret is the ring-shaped stand the gun mount is mounted on. An oblong door leads into the upper ammunition-handling room below the gun mount. Over that door is a rain gutter in the shape of an inverted V. Four ladder rungs are on the side of the shield toward the top; other rungs below them have been removed, but the mounting holes for the screws that held the rungs in place are still apparent.

The rear of the shield of 5-inch gun mount three is displayed. The small door with the rounded bottom toward the bottom of the shield is for the case-ejection chute port. The small square door above it is for the auxiliary case-ejection port.

On the starboard side of the superstructure deck among the 5-inch/38-caliber gun mounts is a cable reel, for storing cables. On the side of the drum is a ring-gear and pinion assembly, for operating the reel with a hand crank (not installed).

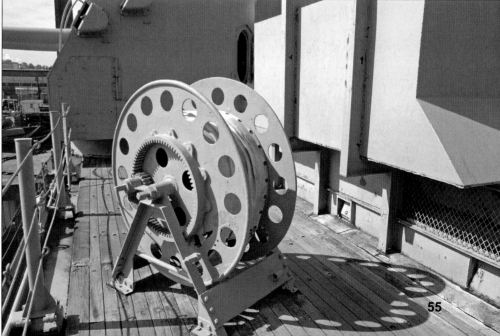

The cable reel shown in the preceding photo is viewed from another angle, facing forward on the superstructure deck. A canvas cover is lashed around the stored cable on the reel, to protect the cable from the elements. Numerous cable reels were on the decks.

Blast bags, or bloomers, when present, were fastened to the rectangular frames mounted around the gun ports. The curved hoops are bloomer protection rails, intended to keep the ample folds of the bloomers from becoming fouled in the gun ports.

The upper ammunition-handling room below 5-inch/38-caliber gun mount five is viewed, with the entry/exit door to the right. The structure is of welded construction. At the lower corner of the structure is a ventilator screen. To the left is the adjacent cable reel.

On the superstructure immediately above the top of the shield of 5-inch gun mount five are the foundation and tower for the starboard center Mk. 37 secondary-battery director. The round platform to the right of the gun mount supports a quad 40mm Bofors anti-aircraft gun mount.

The upper ammunition-handling room of 5-inch/38-caliber gun mount number five is displayed, with the forward part of the ship to the right. The part of the structure in the foreground is a ventilator housing incorporated into the outboard side of the structure.

Inboard of 5-inch gun mount seven, mounted on the aft side of the ammunition-handling room of 5-inch gun mount five, are four steel lockers that at least one source identifies as 20mm ammunition lockers, but there were no 20mm gun mounts in this immediate area.

Just aft of 5-inch gun mount five is 5-inch mount seven. Like 5-inch mount three, as well as mounts four and eight on the port side of the superstructure deck, mount seven's stand rests directly on the deck, and its upper ammunition-handling room is one level below.

The structure containing the upper ammunition-handling room for 5-inch/38-caliber gun mount number nine is viewed facing aft. On top of it is the ring-type stand for the gun mount, which contains the roller bearings upon which the mount rotates.

The aftermost 5-inch/38-caliber gun mount, number nine, is positioned above the structure containing the upper ammunition-handling room. To the left is the rear of 5-inch/38-caliber gun mount number seven, and to the right are the boat crane and a boat.

The *Massachusetts* had several ship's bells; this one is on the starboard side of the superstructure deck aft of 5-inch gun mount nine. Shrapnel damage caused by an enemy 8-inch shell during the Battle of Casablanca on 8 November 1942 is indicated.

A quad 40mm gun mount and 5-inch/38-caliber gun mounts seven and nine are in the foreground in this view facing forward on the starboard side of the main deck toward the rear of the superstructure. To the left is the kingpost of the boat crane, with part of the boom of the crane being visible between mount nine and the mainmast. The lower part of the mainmast has diagonal braces on each side, and ladders running up the mast are faintly visible. Braces securing the upper extent of the foremast to the forward fire-control tower are also in view.

Some of the features on the aft starboard side of the superstructure are displayed, including the smokestack (right), 5-inch/38-caliber gun mounts seven and nine, and the boat crane located on level 2, one level above the superstructure deck. On the side of the superstructure one level down from the boat crane are ventilators; a ship's bell is hanging to the right of them. The pipes with flared tops running along the rear of the smokestack are steam-escape pipes. Inboard of the boat crane kingpost is the aft fire-control tower, on top of which is the aft Mk. 38 primary-battery director, nicknamed "Spot 2." A searchlight is mounted on a small platform with guard rails on the rear of the aft fire-control tower.

As observed from aside turret three facing forward, the USS *Massachusetts*' masts and directors bristle with radar antennas. In addition to these, there were various radio antennas and IFF (identification friend or foe) antennas on the ship.

On the roof of 5-inch/38-caliber gun mount ten, toward the aft port terminus of the superstructure deck, is a blast hood over the mount captain's hatch, to protect him from concussion from the blasts of nearby guns when he was observing through the hatch.

Inside a gallery in the port side of the superstructure on level 2 is a twin 5-inch loading machine. This was an apparatus upon which 5-inch gun crews could hone their proficiency at loading the guns without being confined within the gun mounts.

Looking aft from the port side of the superstructure deck, with a 5-inch gun mount to the right, the aft Mk. 38 primary-battery director, "Spot 2," looms above the superstructure. On South Dakota-class battleships, different versions of the primary-battery directors were installed in the forward and aft positions: the forward director was the Mod. 2 and the aft one was the Mod. 3. Normally, the personnel assigned to a Mk. 38 director were a spotter, rangefinder operator, standby rangefinder operator and talker, pointer, trainer, cross-leveler, and radio operator. In addition to a number of telescopes and a 26.5-foot stereoscopic rangefinder, the director was equipped with several successive types of radar; currently mounted on the aft Mk. 38 director is a Mk. 8 radar antenna.

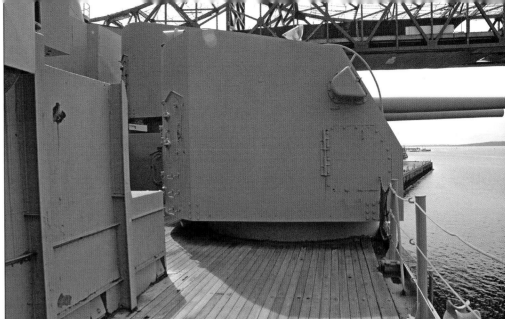

The 5-inch guns of mount eight are trained over Battleship Cove. Partially visible aft of and higher than that mount is mount ten. Protruding from the front of the shield is one of the bloomer protection rails that kept the bloomers from getting snagged in the gun ports.

The rear and roof of the shield of a 5-inch/38-caliber gun mount on the port side of the ship are displayed. The roof-hatch operating mechanism has been removed from the upper center rear of the shield. The spotlight on the roof is a modern addition.

The bracket to which the bottom of the roof-hatch operating mechanism was attached is still on the rear of the shield. The cutout in the bottom center of the blast hood provided clearance for the connection between the hatch operating mechanism and the hatch.

Several hawser-stowage reels and their mounting brackets on the port side of the superstructure deck are viewed from above. To the right is a storage locker, and at the top of the photograph is the rear of a 5-inch/38-caliber gun mount.

The three hawser-stowage reels displayed in the preceding photograph are viewed from the superstructure deck, showing the brackets that secure them in an upright position. Two of the reels have protective canvas wrappings lashed in place around the cables, to protect them from moisture. The area is slightly forward of the smokestack; the two port steam-escape pipes on the rear of the stack are visible to the upper right. Above the two cable reels on the right, the vertical tube and slanting structure above it are part of the foundation for the port center Mk. 37 secondary-battery director, which is out of the view above the top of the photo.

The quad 40mm antiaircraft mount indicated in the preceding photo is viewed close-up facing aft. The trainer's and the pointer's control hand wheels are present on the mount, as are the ring sights with their guards. Five-inch gun mount two is in the background.

Mounted on the superstructure deck near the front of the superstructure is a whip antenna with an expanded-steel mesh guard around its lower part. In the background are the left rear of turret two (left) and the front of the conning tower, behind the antenna.

The forward part of the superstructure is seen from the port side of the main deck. The door at the bottom leads into a corridor providing access to several wardroom staterooms, a fan room, and other compartments. Aft of the stairway landing at the center, on the superstructure deck, is a quad 40mm antiaircraft gun mount mounted in a tub, just aft of which is 5-inch/38-caliber gun mount number two. To the upper left is the bridge, and above it is the top of the conning tower, on which one of the vision slots is visible. Rising above the superstructure is the forward fire-control tower, with the Mk. 38 Mod. 2 primary-battery director on top of it.

A ventilator plenum is in the foreground of this image of the left side of turret two facing forward, with the left 16-inch gun of turret one visible in the distance. At the top center is the left hood of the rangefinder. A handrail is mounted under the top of the plenum.

From approximately the same vantage point as in the preceding photo, the rear of turret two is shown, including the platform with guard rails suspended under the gun house hatch and the handrails and foot rails that were retrofitted at some point.

The conning tower at the superstructure-deck level is viewed from a different angle than that shown in the preceding photograph. The prominent indentation in the conning tower is present on both sides of the structure. On this part of the conning tower, the armored structure is clad with a metal skin secured with rivets and welded seams. Also present on the tower are electrical lines and a wire-antenna fitting. Toward the top, the struts that help stabilize and strengthen the bridge are in view. Bracing strips also were applied to the side and bottom of the bridge.

The left hood of the rangefinder of turret two is viewed close-up, showing the access plate screwed to the side of the hood and the viewing port on the front of the hood. The hood is attached to the gun house with hex screws. At the top right is the bridge.

The hoods for the left trainer's (upper) and pointer's telescopes protrude from the side of turret two. There was a pointer and a trainer on each side of the gun house, and their duty was to lay the guns on the target when the mount was under local control.

The front port area of the superstructure is observed from the superstructure deck, with two guns of the quad 40mm gun mount appearing in the foreground. The left wing of the bridge is at the upper left. At the center of the photo is a round platform and tub for another quad 40mm Bofors antiaircraft gun mount; the gun barrels are peeking over the top of the tub. To the bottom right is the right side of 5-inch/38-caliber gun mount number two; directly above it is the port center Mk. 37 secondary-battery director, mounting on top of it a Mk. 12 radar antenna combined with a Mk. 22 "orange peel" radar antenna.

The port side of the bridge of USS *Massachusetts* is in view, including the enclosed section at the front of the bridge with square windows. A board with campaign ribbons is also present. A good view is provided of the system of struts underneath the bridge.

Above the superstructure at the top right, the left side of the forward Mk. 37 secondary-battery director, "Sky 1," is visible. The box-shaped object near the top of that side is a cover where originally the cylindrical-shaped left side of the rangefinder protruded.

Underneath the bridge is a network of frame members that are continuations of the exposed braces on the sides of the bridge above. Also underneath the bridge are exposed electrical cables for the navigational equipment located on the bridge.

At the center of the photo, on the port side of the level atop the navigating bridge, known as the house-top, is a curved splinter shield. The raised enclosure at the rear of the curved splinter shield is a tub housing a Mk. 51 director. The starboard side is similar.

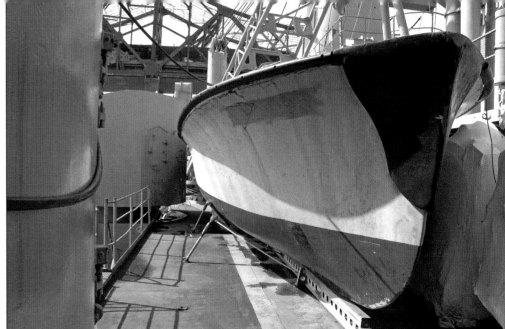

A covered motorboat is stowed on the starboard side of level 2, one level above the superstructure deck, alongside the rear of the smokestack. As World War II progressed, the number of boats carried onboard the ship was reduced.

The motorboat is viewed from off its starboard bow. It rests on chocks, with outrigger braces to stabilize the boat. This method of stowing the boat saved on weight. Portions of the boat crane's boom and kingpost are visible aft of the boat.

A close-up shows details of the chocks and the outrigger braces for the motorboat. In the foreground, a stabilizer rod is connected to the brace in the foreground and to the chock's beam. To the left is the rear of 5-inch gun mount nine.

The *Massachusetts*' boat is observed from aft, showing its single propeller and rudder. Visible above the boat and projecting from the side of the superstructure are two small antiaircraft gun-director tubs.

Most of the boat crane is visible in this view from off the starboard beam of the ship. The kingpost is adjacent to the rear of the aft fire-control tower, while the boom extends at an angle behind the 5-inch/38-caliber gun mount. Originally the ship had two cranes, but now only this crane remains. The crane rotated around the kingpost. The heel of the crane is attached to pivot points on a platform around the crane; this platform also holds the hoisting gear and cable drums. Other platforms around the kingpost were dedicated to the topping gear and reduction gear. The maximum elevated angle of the boom under a full load was 70 degrees from horizontal.

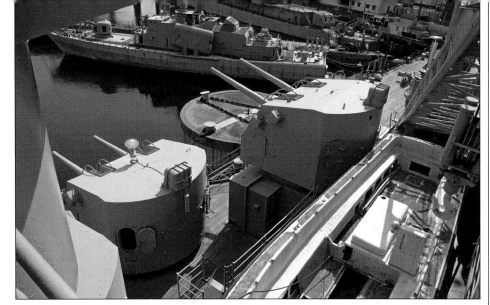

Part of the interior of the boat is viewed from above. In the background are two display ships at Battleship Cove: the Russian-built *Hiddensee,* a *Tarantul I*-class corvette, and, farthest away, the *Joseph P. Kennedy Jr.* (DD-850), a *Gearing*-class destroyer.

Specifications and General Data

Length Overall	680' 0"
Length at waterline	666' 0"
Maximum beam	107' 11"
Waterline beam	108' 2"
Mean draft	31' 7 5/16" @ 42,329 tons
Maximum draft	34' 9½" Forward, 36' ¾" aft
Displacement	35,113 tons Light 1946 45,216 tons Full Load 1946
Machinery	Boilers: eight Babcock and Wilcox Geared turbines: four sets General Electric, 133,000 shaft horsepower forward, 32,000 astern
Speed	27.08 knots
Complement 1945	2,339 (168 officers, 2,500 enlisted)
Cost	$76,885,750.00
Laid down	20 July 1939
Launched	23 September 1941
Commissioned	12 May 1942

The boat crane is observed from the forward end of the boom. The yellow hook hovering above the bow of the boat is the boat hook, which had a capacity of 27,000 pounds. The maximum extension of the boat hook from the center of the kingpost was a little over 52 feet. The crane and boom also accommodated a cargo whip, or cable, routed over the sheave at the very end of the boom, and attached to the end of the whip is the whip hook, seen at the bottom right. This hook was rated at 7,000 pounds. To the right is the rear of the smokestack, with the two starboard steam-escape pipes running along it.

The boat hook is mounted on a block, and the boom sheaves over which the boat hook cables pass on their way back to the hoist winch are mounted several feet back from the end of the boom. The sheave at the end of the boom accommodates the whip cable.

The hoisting-gear platform is viewed facing aft, showing the mounting points for the heel of the boom. The wide drum is for the boat-hook cable, while the narrow drum to the right is for the whip cable. To each side of the drums is a winch gearbox.

As viewed from the rear of the boat crane on level 2, the control platform (top) and hoisting-gear platform (bottom) surround the kingpost. Visible on the control platform is an array of hand wheels for operating the crane. On the platform below it, there is a hydraulic motor on each side of the kingpost, to power the hoisting winches. The lower part of the kingpost extends below this level, where there are two more platforms with crane machinery, including topping machinery, which controlled the angle of the boom, and the reduction gear for the electro-hydraulic power drive of the crane. The topping, hoisting, and rotating gears each had its own separate power source.

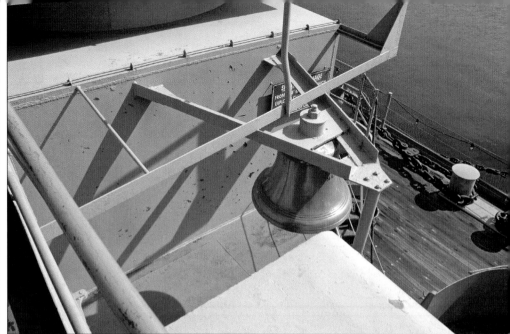

The boat crane is observed obliquely from the rear, showing the topping sheaves and cables that operated the boom. A ladder is attached to the rear of the kingpost. To the top left are the aft Mk. 38 primary-battery director and a platform with a 24-inch searchlight.

A few steps to the starboard of the kingpost of the boat crane on level 2 is a railing with a view down to the ship's bell and the shrapnel damage from the November 1942 Battle of Casablanca. The steel deck below the bell is the superstructure deck.

Toward the aft end of level 2 are two side-by-side quad 40mm gun mounts. The one in the foreground is the starboard mount, while the port mount is to the right. To the far left is the quad 40mm gun mount atop turret three. The aft Mk. 37 director is at the center.

Features on the aft portion of level 2 are observed from the starboard side, with the rear of turret three to the left. The quad 40mm gun mount seen in the preceding photo is near the center of this photo, just forward of the aft Mk. 37 secondary-battery director.

The rear of the aft Mk. 37 secondary-battery director is observed on level 2. In the background, the Braga Memorial Bridge over the Taunton River poses a challenge to discerning the framework supporting the Mk. 12 and Mk. 22 radar antennas.

Directly below the housing of the aft Mk. 37 director is its ring-type stand, with exposed electrical conduits, cables, and boxes present on it. Some of the electrical components have been stripped, leaving exposed several brackets and mounting holes.

The Mk. 12 radar antenna atop the aft Mk. 37 secondary-battery director is viewed from the starboard stide. The Mk. 12 was a 33cm radar capable of automatically tracking the range of the target and calculating its range rate at ranges of up to 45,000 yards.

As seen from the right rear quarter of the aft Mk. 37 director, the large antenna at the center is the Mk. 12, while the oblong, parabolic antenna to the right is the Mk. 22, which calculated the height of an approaching aircraft. It measures 1.5 by 6 feet.

In the foreground is the aft starboard quad 40mm antiaircraft gun mount near the rear of the superstructure on level 2, showing the trainer's ring sight and control hand wheel. The tub containing the director that controlled this quad 40mm gun mount is at the top center. The view is facing forward, with the kingpost of the boat crane in the center background. Directly above the 40mm gun barrels is the platform surrounding the kingpost that contained the controls for the boat crane. Below the director tub is a loudspeaker, and mounted on the side of the tub is a spotlight.

Looming directly above the aft port quad 40mm gun mount near the rear of level 2 is the tub for the director that controlled that mount. Visible above the top of the tub is a Mk. 57 director. This director, introduced late in the war, was equipped with a Mk. 34 radar.

The forward ends of the tubs of the quad 40mm gun mounts toward the rear of level 2 are open. This is the port mount. The stanchion in the foreground supports a gun director and tub above. At the upper center is the aft Mk. 37 secondary-battery director.

On the port side of level 2 adjacent to the rear of the superstructure is a 40mm loading machine, an apparatus upon which 40mm gun crews could practice their proficiency and speed in loading their weapons without having to do so on the actual weapons.

The 40mm loading machine is viewed from the side. Crewmen loaded four-round clips of dummy ammunition into the replica of an automatic loader, the black fixture atop the device to the right. The machine is powered by electric motors driving flywheels.

As viewed from its rear, the 40mm loading machine has two facsimiles of an automatic loader, the mechanized feed chute on top of a 40mm Bofors gun's receiver. This feature of the loading machine mimics the arrangement of the actual 40mm gun mount, where two guns are mounted together. Thus, two gun loaders can stand to each side of the loading machine and practice loading ammunition clips. On the right of the loading machine is the right flywheel with a guard cage over it. The electric motor that drove this flywheel is mounted under the front end of the loading machine.

The rear of the shield of one of the 5-inch/38-caliber gun mounts on the port side of the ship is viewed from a level even with the roof of the mount, illustrating details of the blast hood, main and auxiliary spent-casing ejector doors, and crew access doors.

The two fixtures between the 5-inch loading machines are projectile hoists. In the actual 5-inch gun mounts, a projectile hoist was an apparatus that conveyed 5-inch projectiles by electro-hydraulic power from the below-decks ammunition-handling room up to the gun house.

On level 2 between frames 93 and 97 is a gallery containing two 5-inch/38-caliber loading machines. Similar to the 40mm loading machine, these fixtures enabled gun crews to practice loading their pieces, except out in the open air, unconfined to a gun house.

The fixtures resembling a box with a drum to the rear atop the loading machines are the powered rammers, which rammed the projectiles and powder charges into the chamber. The boxy structure is the rammer tank, and the drum-shaped object is the rammer motor.

Level 3, the third level above the main deck, is the flag bridge. The view is to the port quarter. On the left is the flag bag, where the ship's signal flags were methodically stored. On the opposite side of level 3 is the starboard flag bag, similar in layout to this one.

Details of the bulwark and its framing on the port side of the flag bridge are visible. To the top left is the blast hood over the mount captain's hatch on 5-inch gun mount two. In the background is a platform with guardrails for a 24-inch searchlight.

The full extent of the curved bulwark shown in the preceding photo is shown, with the port flag bag to the extreme left. As can be seen, this curved bulwark, along with a corresponding curve in the deck, provided clearance for the 5-inch gun mount two.

The 24-inch searchlight on the port side of the flag bridge is seen up close, showing the clear lens on its front. The lens is installed on a hinged, ring-shaped frame with a grab handle at the front, and the lens and frame are secured in place by clamp handles.

Tubular legs with diagonal braces support the platform for the 24-inch searchlight on the flag bridge. The door with a window in the background leads into the port side of the forward, enclosed portion of the flag bridge. To the upper right is the navigating bridge.

A view of the 24-inch searchlight on the port side of the flag bridge shows the ventilating-fan housing on top, to the right of which is an open sight and a bracket for a spotting telescope. Inside the lamp was an iris shutter for turning off the beam.

The 24-inch searchlight on the port side of the flag bridge is viewed from above facing aft, showing the circular shape of the platform. Illumination of the searchlight was by carbon arc, and the searchlight was used primarily for signaling, with secondary uses for navigating and visual searching and detection. Atop the barrel of the light is a ventilating-fan housing. According to deck plans for the USS *Massachusetts*, originally an admiral's chair was mounted on the deck just forward of the searchlight platform, to the bottom of the photo. The barrels of all five of the port 5-inch/38-caliber gun mounts are visible in this view, giving a sense of the tremendous firepower these weapons could bring to bear on enemy targets in the air or on the surface.

Near the door into the enclosed part of the starboard side of the flag bridge is an alidade mounted on a D-shaped platform. When the dome-shaped cover of the alidade was removed, it exposed a sighting device and azimuth scale, for establishing bearings.

The 24-inch searchlight on the starboard side of the flag bridge is observed facing forward. The round device aft of the grab handle on the frame of the lens is the signal-shutter operating motor. A handle would be installed on this motor to tap out signals.

A fine view is offered of the forward part of the main deck and turrets one and two from the forward starboard corner of the open-air portion of the flag bridge. The bulwark was designed with an outward curve at the top to deflect wind and water; several braces on the inner side of the bulwark are in view. To the far left is a portion of an alidade, an instrument that enabled the operator to take bearings on distant objects and landmarks. At the top left is part of the navigating bridge, showing some of the electrical wiring and framing underneath it.

The starboard 24-inch searchlight and platform on the flag bridge are viewed facing aft. The box-shaped assembly on the bottom of the drum of the searchlight is the lamp housing. The handles on the rear of the drum were for manually training and elevating the searchlight. The searchlight was of the automatic high-intensity carbon-arc type and had unlimited rotation and a range of elevation from from 110 degrees to -30 degrees. To the upper right is the underside of the navigating bridge deck, and the rounded platform and tub projecting above the searchlight is a quad 40mm antiaircraft gun mount on the house top: that is, the level above the navigating bridge.

Aft of the searchlight on the starboard side of the flag bridge, a view of the starboard center Mk. 37 secondary-battery director is available. Visible above the starboard flag bag at the bottom of the photo is the cylindrical foundation post for the director. The angled structure above that post is also part of the support for the director as well as the section of the house top that the director is mounted on. A similar structural arrangement is also present on the port side for the center port Mk. 37 director. The deck with the guard rails to the lower right is the navigating bridge deck. A glimpse of the upper part of the mainmast is at the top right.

From alongside the curve in the bulwark on the starboard side of the flag bridge designed to give clearance to 5-inch/38-caliber gun mount number one, the angled braces and the stanchions that support the 40mm gun mount on the house top are in the foreground.

The starboard flag bag is on the flag bridge aft of the curved section of bulwark. To transmit signals by flags, flags stored in the flag bag would be run up halyards attached to yardarms extending from the upper part of the forward fire-control tower.

Ship Signal Flags

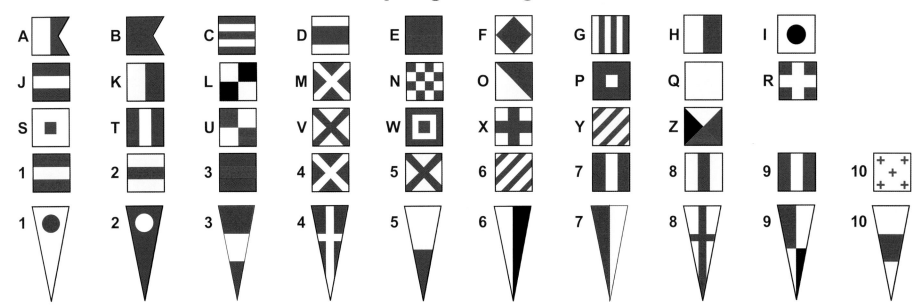

The bridge (lower right) originally was unenclosed, with a narrow walkway hugging the conning tower. In mid-1944 an enlarged but still open bridge, with straight sides with vertical bracing, was installed. The present, enclosed, bridge dates to January 1945.

Armor Specifications

Main battery turrets	Face plates: 18.0"
	Sides: 9.5"
	Back: 12.0"
	Roof: 7.25"
Barbette Armor	Centerline: 11.6"
	Sides: 17.3"
Secondary battery mounts	2.0" 78-lb Special Treatment Steel (STS)
Magazines	2.0"
Conning tower armor, Class B	Sides 16.0"
	Beam: 16.0"
	Roof: 7.25"
	Communications tube: 16.0"
Belt armor	12.2" Class A on .875 30-lb STS inclined 19 degrees
Lower belt armor	12.2" Class A on .875" 30-lb STS tapered to 1" on .75
Deck armor	Main: 1.5" (centerline and outboard)
	Second: 5.0"+7.5" (centerline) - 5.3"+.75" (outboard)
	Third: 0.3" (centerline and outboard)

On the house top, the level atop the navigating bridge, is a splinter shield (center) enclosing two Mk. 57 directors and, in raised tubs to the rear of the shield, two Mk. 51 directors. The tall object to the left of the splinter shield is a Mk. 27 radar antenna.

The starboard side of the navigating bridge is viewed from the enclosed area at the front of the navigating bridge. In the foreground is the conning tower, including one of its vision slots, showing the great thickness of the armor on this structure.

An alidade located on the starboard side of the navigating bridge, visible in the background in the photo at left, is seen here up close. The dome-shaped cover of the alidade has been removed, exposing to view the azimuth scale housed in the top of the alidade.

The starboard side of the navigating bridge is viewed facing aft from the inboard side of the alidade (which is outside of the view to the left). Projecting from the house-top level is a quad 40mm antiaircraft gun mount and tub, below which are several 5-inch gun mounts.

From the same vantage point as in the preceding photo, a good view is available of the 24-inch searchlight and platform on the starboard side of the flag bridge. All five of the 5-inch/38-caliber gun mounts on the starboard side of the ship are visible.

Facing forward from the starboard side of the house top, the starboard side of the navigating bridge is below to the right (note the alidade in the jog in the bulwark), the top of the conning tower is to the top left, and a gun-director tub is at the top center.

A view of from the port side of the house top facing forward depicts the port side of the navigating bridge. The two small "boxes" on the rounded bulwark at the top are radio antenna trunks, where leads from wire antennas pass through insulators and into the ship.

Protruding through the top of the conning tower are five periscopes, which the personnel manning the tower relied on for visibility, since the tower's direct-vision fixtures were limited to a relatively few vision slots. The radar antenna is a non-original fixture.

On the starboard side of the house top aft of the conning tower (right) is a Mk. 57 director (center). Another Mk. 57 director is on the port side of this level. To the right is a raised tub for a Mk. 51 antiaircraft director, with ladder rungs providing access to it.

On the rear of the top of the conning tower is a Mk. 27 standby fire-control radar antenna (left of center) on a box-type mount. This radar system replaced the Mk. 40 stereoscopic rangefinder that was originally located on top of the conning tower.

Nestled between the conning tower (right, with the Mk. 27 radar mount at the top) and tub for a Mk. 51 director (left) is the Mk. 57 director on the port side of the house top. The radar antenna of the Mk. 57 director is on the front of the unit, facing toward shore.

The Mk. 51 director tub aft of the Mk. 57 director on the port side of the house top is viewed. The top of the director is visible above the top of the tub, but the gun sight is not installed. Some of the original electrical boxes and controls are still on the tub.

On the front of the forward Mk. 37 director are hinged port covers for the telescopes of, left to right, the trainer, the pointer, and the control officer. Foot rails and hand rails are arranged around the housing to enable crewmen to access exterior components.

The forward Mk. 37 secondary-battery director, nicknamed "Sky 1," is viewed from behind, with the rear of the conning tower visible to the far right. On the right side of the director is the box-shaped protective housing installed over the right side of the Mk. 42 rangefinder at some point after the ship's 1945-1946 refit: probably when the ship was being prepared for long-term storage with the Atlantic Reserve Fleet. A similar housing is over the left side of the rangefinder. On the rear of the cylindrical foundation of the director is a crew access door, above which is an outrigger to which are attached wire antennas.

At the lower center is the housing over the left extension of the Mk. 42 rangefinder of the forward Mk. 37 director. A small vision port is in the front of the housing, apparently to permit checking the operation of the optics of the rangefinder. Mounted on the side of the director to the front of the rangefinder housing is a vertical ladder. To the top right, part of the Mk. 12 radar antenna is visible. In the background is the forward fire-control tower. Visible on the front of the tower just above the top of the Mk. 37 director is a 36-inch searchlight on a platform. The platform and bulwark surrounding the conning tower two levels above the searchlight are part of the forward air-defense station.

The right housing for the Mk. 42 rangefinder on the forward Mk. 37 director lacks the viewing port that is present on the left housing. Above that housing, to the side of the Mk. 12 radar antenna, is the Mk. 22 "orange-peel" radar antenna. This radar unit enabled the Mk. 37 director to accurately gauge the height of an oncoming enemy aircraft, thus resulting in a much improved firing solution than was possible before the Mk. 22 radar was added to the director. At the top of the forward fire-control tower is the Mk. 38 main-battery director, with the rangefinder housings extending from each side of the director. To the bottom right is the left side of the center starboard Mk. 37 director.

On each side of the house top, just aft of the forward Mk. 37 director, is a platform and tub for a quad 40mm antiaircraft gun mount. This one is on the port side. The tub is fitted with foot rails and handrails, and two diagonal braces help support the structure.

The quad 40mm gun mount on the port side of the house top has been stripped of some of its components, including the ring sights and sight guards, as well as the ready-ammunition racks that originally lined the inside of the gun tub.

The trainer's footrests are still present on the quad 40mm gun mount on the port side of the house top, but his seat and hand controls are missing. In the background is the center port Mk. 37 secondary-battery director and its cylinder-shaped foundation. Optical and radar information from this director and the other three Mk. 37 directors was routed to the plotting room below decks, where analog computers calculated firing solutions for the 5-inch/38-caliber guns. The Mk. 37 directors could also be used to control the 40mm guns and, in emergency conditions, the 16-inch guns.

The quad 40mm gun mount on the starboard side of the house top is viewed facing aft. Components are also missing from this mount, including, as seen here, the pointer's ring sight, seat, and hand controls. The guard rail is also missing from the loaders' platform at the rear of the mount. Details are visible of the handrail and its mounting brackets around the upper exterior of the gun tub. The starboard center Mk. 37 secondary-battery director is present in the background. The drum-shaped structure to the upper right is a Mk. 51 director tub; a similar one is on the opposite side of this level.

On each side of the aft end of the house-top level, adjacent to the smokestack, are two Mk. 51 directors in tubs. This one is on the starboard side, facing aft, with the tub of another Mk. 51 director partly visible to the left. Full-length openings in the tubs provided easy access to the director operator. The sight and several other components are no longer mounted on the director. The smokestack is to the right of the directors, just outside of the photograph. In the background to the lower right is part of the roof and the rear of the shield of 5-inch/38-caliber gun mount number nine.

In a view from the starboard beam, the two Mk. 51 directors seen in the preceding photograph are next to the smokestack, just aft of the foundation of the center starboard Mk. 37 director. Many of the ship's radar and radio antennas are also displayed.

The smokestack is viewed looking upward from the starboard side. The tubs of the two Mk. 51 directors indicated in the preceding photos are in the foreground. At the top of the stack is the funnel cap, painted matte black; whip antenna supports are on its side.

The foremast (left) and mainmast (right) are portrayed in this view of the smokestack from the port quarter. A Mk. 51 director and tub was mounted directly to each side of the smokestack in a July 1944 refit, but these were removed during a 1946 refit.

The port side of the flag bridge is viewed from alongside the smokestack facing aft, with the angled struts of the mainmast and the aft Mk. 38 main-battery director appearing in the left background. There is no corresponding section of flag bridge on the starboard side, in order to provide clearance for the boat crane. As built, there was a boat crane and boat stowage in the area depicted in this view, but these were eliminated during a refit in World War II. During the war, there was a gallery of five 20mm Oerlikon antiaircraft gun mounts along this part of the flag bridge.

Looking forward from the aft port part of the flag bridge, the mainmast and its struts are in the foreground. The two Mk. 51 director tubs at the aft port corner of the house-top level in the background have been touched-up with red primer. Just beyond these two tubs is the center port Mk. 37 secondary-battery director, pointing outboard. Outriggers jut from the mainmast at the level of the funnel cap, and the outriggers are stiffened by diagonal braces. A small searchlight is mounted on a stanchion adjacent to the port brace of the outriggers.

The aft part of the superstructure is viewed from the starboard side, with the smokestack positioned between the mainmast and the foremast. The kingpost of the boat boom is toward the left, to the upper left of which is the aft Mk. 38 main-battery director.

The funnel cap, whip antenna mounts, foremast, and, to the right, the center starboard Mk. 37 director are observed. The large dish antenna on the foremast at the center is the SK-3, above which on the top foremast is an SG-1B surface-search radar antenna.

Radar antennas on the mainmast (left) and foremast are seen from the aft starboard quarter. SG-1B surface-search radar antennas are near the tops of both masts. The large, rectangular antenna on the maintop is the SR-A secondary air-search radar.

The aft Mk. 38 main-battery director, "Spot 2," is observed from near the aft port quarter of the smokestack (far left) facing aft. This director is equipped with a Mk. 8 fire-control radar antenna on top, whereas the forward Mk. 38 director has the Mk. 18 radar.

Another view of the aft Mk. 38 main-battery director shows it from the starboard side. The director is mounted on a funnel-shaped tower, with a searchlight platform (partially obscured by the kingpost of the boat crane) partway up its aft facet. The director on top of the tower rotated to track enemy surface and land targets. The 26.5-foot stereoscopic rangefinder protruding from each side of the director provided optical range information, while the 10cm Mk. 8 fire-control radar could quickly fix the range of an enemy target. The polyrods protruding from the front of this antenna gave this antenna its nickname, the comb antenna.

The front of the Mk. 8 fire-control antenna atop the aft Mk. 38 main-battery director is displayed. This radar was manufactured by Western Electric and is considered the first track-while-scanning radar. It featured a wide field of view and could pinpoint the range of a capital ship at 40,000 yards or a submarine at 10,000 yards. While the Mk. 8 radar on the forward Mk. 38 director was replaced by the Mk. 15 radar during a 1945-1946 refit, the Mk. 8 radar remained on the aft Mk. 38 director. Below the Mk. 8 radar antenna is the searchlight mounted on the aft facet of the aft fire-control tower.

The rotating part of the Mk. 38 director is above the searchlight. At the outer ends, hinged covers protect the openings for the rangefinder objectives. On the front of the center part of the director are hinged covers for the trainer's and pointer's telescopes.

A Mk. 57 director equipped with radar is mounted in a tub to the rear of the aft fire-control tower. The radar antenna is to the right of the director's pedestal. Below are turret three and the aft part of the main deck. To the bottom right is a quad 40mm gun mount.

Inside the gun house of a 5-inch/38-caliber gun mount, the rear of the right gun is to the right, and the rear of the left gun is to the center. The left crew door is open to the left. At the top right is the rammer motor. In the background is a ventilator duct.

Each of the 10 5-inch/38-caliber gun mounts has a corresponding powder-handling room below decks, where powder charges for the guns are brought in from magazines and sent up hoists to the gun mounts. Shown here are two hoists with their doors open.

At the rear of each 16-inch gun house is the turret officer's booth; the view faces the right side. To the right is the rangefinder, and a yellow periscope is to the left. Rammers for the guns are on the floor. The gun chambers are on the other side of the bulkhead to the left.

The ship's eight Babcock & Wilcox boilers produced steam that powered four sets of General Electric geared turbines. The General Electric double reduction gears, the housing for one of which is shown here, converted the turbine rpm to rpm suitable for the propellers.

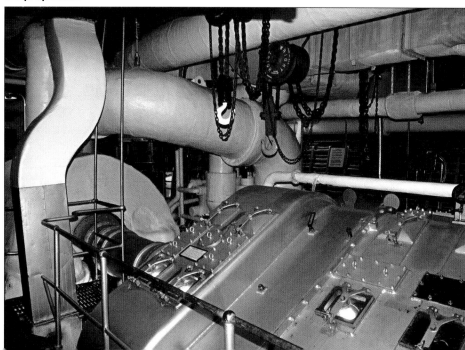

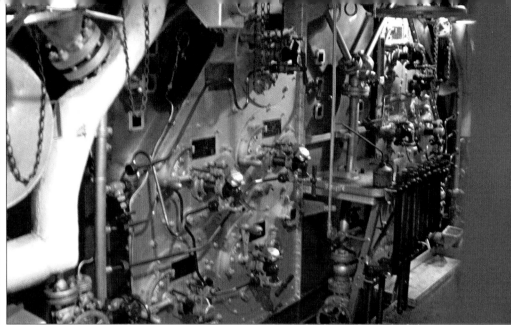

The main galley is where most of the cooking for the ship's company was done. It is on deck 2 and abuts the barbette of turret three, which is visible behind the soup kettles. Also in view are sinks, food preparation tables, ventilation trunks, and exposed plumbing.

One of the Babcock and Wilcox three-drum express boilers is displayed. It is a twin-furnace design with a working pressure of 578 psi and temperature of 850 degrees Fahrenheit. To conserve space, designers of the *South Dakota*-class battleships grouped the boilers in the engine rooms with the turbines and reduction gears.

Radio central, on the third deck, is the place where operators sat at their sets, day and night, sending signals and receiving and typing transcripts of messages and orders from the fleet commands. Transcripts of coded messages were then sent to the decoding room.

The engine room power was transmitted to the water via these four massive propellers. The five-bladed outboard props are 17' 4½" in diameter, while the inboard three-bladed props measure 17' 8". Two semi-balanced rudders guided the ship. (Moe Knox via Battleship Cove)

Today, USS *Massachusetts* forms the centerpiece of Battleship Cove, in Fall River, Massachusetts. Her big guns, long silent, provide mute testament to a bygone day of naval warfare. Inner spaces that were once the home to thousands of fighting men now form a memorial to those men, and hundreds of thousands more like them, who manned vessels like these, protecting the shores of the United States, and fighting for justice around the globe. (Battleship Cove)